Shenandoah Retirement Home Fire
Roanoke County, Virginia

Investigated by: Randolph E. Kirby

This is Report 038 of the Major Fires Investigation Project conducted by TriData Corporation under contract EMW-88-C-2649 to the United States Fire Administration, Federal Emergency Management Agency.

 FEMA

Department of Homeland Security
United States Fire Administration
National Fire Data Center

U.S. Fire Administration Fire Investigations Program

The U.S. Fire Administration develops reports on selected major fires throughout the country. The fires usually involve multiple deaths or a large loss of property. But the primary criterion for deciding to do a report is whether it will result in significant "lessons learned." In some cases these lessons bring to light new knowledge about fire--the effect of building construction or contents, human behavior in fire, etc. In other cases, the lessons are not new but are serious enough to highlight once again, with yet another fire tragedy report. In some cases, special reports are developed to discuss events, drills, or new technologies which are of interest to the fire service.

The reports are sent to fire magazines and are distributed at National and Regional fire meetings. The International Association of Fire Chiefs assists the USFA in disseminating the findings throughout the fire service. On a continuing basis the reports are available on request from the USFA; announcements of their availability are published widely in fire journals and newsletters.

This body of work provides detailed information on the nature of the fire problem for policymakers who must decide on allocations of resources between fire and other pressing problems, and within the fire service to improve codes and code enforcement, training, public fire education, building technology, and other related areas.

The Fire Administration, which has no regulatory authority, sends an experienced fire investigator into a community after a major incident only after having conferred with the local fire authorities to insure that the assistance and presence of the USFA would be supportive and would in no way interfere with any review of the incident they are themselves conducting. The intent is not to arrive during the event or even immediately after, but rather after the dust settles, so that a complete and objective review of all the important aspects of the incident can be made. Local authorities review the USFA's report while it is in draft. The USFA investigator or team is available to local authorities should they wish to request technical assistance for their own investigation.

This report and its recommendations were developed by USFA staff and by TriData Corporation, Arlington, Virginia, its staff and consultants, who are under contract to assist the USFA in carrying out the Fire Reports Program.

The USFA appreciates the cooperation and assistance received from Fire Chief T. C. Fuqua and Fire Marshal Ken Sharp of the Roanoke County Fire and Rescue Department, as well as Ottis L. Burgher, Administrator, Shenandoah Homes, Roanoke, Virginia.

For additional copies of this report write to the U.S. Fire Administration, 16825 South Seton Avenue, Emmitsburg, Maryland 21727. The report is available on the Administration's Web site at http://www.usfa.dhs.gov/

U.S. Fire Administration
Mission Statement

As an entity of the Department of Homeland Security, the mission of the USFA is to reduce life and economic losses due to fire and related emergencies, through leadership, advocacy, coordination, and support. We serve the Nation independently, in coordination with other Federal agencies, and in partnership with fire protection and emergency service communities. With a commitment to excellence, we provide public education, training, technology, and data initiatives.

 **FEMA**

TABLE OF CONTENTS

Shenandoah Retirement Home Fire
Roanoke County, Virginia

Local Contacts: T. C. Fuqua, Fire Chief and
Ken Sharp, Fire Marshal
County of Roanoke
Fire and Rescue Department
3568 Peters Creek Road, NW
Roanoke, Virginia 24019
(703) 561-8070
(703) 561-8100

Ottis L. Burgher, Administrator
Shenandoah Homes
5300 Hawthorne Road, NW
Roanoke, Virginia 24012
(703) 362-5412

OVERVIEW

On December 14, 1989, at 0214 hours, a fire alarm was received by the Roanoke County Fire Dispatch Center from the Shenandoah Retirement Home Center. This is a 6-1/2-story masonry building housing 175 elderly persons. Many of the occupants require canes and walkers for mobility. The second floor of this home is licensed by the State as an adult care center for some 48 persons requiring custodial supervision. On December 14, the building was occupied by 135 residents and three staff members. The building actually houses more than this number, but several were visiting family members or were in the hospital.

The first firefighting units arrived 18 minutes after notification of alarm and saw fire coming from a third floor apartment window at the front of the building. Firefighters entering into the building by way of stairwells located at each end of the structure were met by occupants leaving from the upper floors, and with smoke which was already permeating the stairwells. On the third floor they found heat and heavy black smoke from floor to ceiling. Additional fire personnel and equipment had already been alerted by fire dispatch as a result of an on-duty sheriff's deputy who witnessed the fire's progress prior to fire department arrival and radioed this information to the fire dispatch center.

1

SUMMARY OF KEY ISSUES

Issues	Comments
Cause	Believed to be electrical loading of wall receptacles in a third floor room which ignited prefinished wall paneling.
Detection & Reporting	Fire was detected by hallway smoke detector, sounding building alarms and automatically notifying fire department.
Firefighting	Heavy black smoke, generated from burning prefinished wood paneling, permeated entire third floor, making search and rescue extremely difficult and time consuming.
Fire Department Response	Eighteen minutes elapsed between alarm notification and the arrival of first firefighting equipment. This delay probably contributed to extent of smoke and flame damage.
Code Compliance	Building not subject to current code requirements for fire sprinklers and smoke evacuation systems since it was constructed in 1970, prior to such requirements.
Building Structure	Sound masonry and concrete construction prevented serious structural damage. Building was constructed under earthquake building requirements.
Building Occupants	Building is fully occupied by upper age group senior citizens, the majority of whom use canes, walkers, and wheelchairs for mobility, which complicated the evacuation process.
Interior Finish	Apartment units and center core elevator lobby areas on each floor have highly flammable prefinished wood paneling, which produced rapid flame spread and smoke.
Building Evacuation	While residents throughout the building began to evacuate immediately, exit from the third floor was complicated because of heavy, thick black smoke.
Emergency Disaster Plan	Roanoke County's Emergency Disaster Plan for evacuation and shelter management worked almost perfectly and was a major factor in minimizing further injury and possible death to occupants.
Emergency Vehicle	Poor access to the building and fire hydrants required placing apparatus on lawn to reach a fire hydrant and structure, which hampered proper placement of additional apparatus.
Weather Conditions	Frigid temperatures (14 degrees Fahrenheit), snow, and ice-covered roads complicated equipment placement and resident evacuation to transportation vehicles.
High Fire Load & Lack of Fire Inspections	A sampling of apartments revealed high fire loads from storage of furniture and other possessions, and improper use of electrical distribution. Infrequent inspections of apartments by building management, and lack of county authority to inspect them, allowed these conditions to exist.

All tenants were evacuated from the second floor adult care center and were placed in the lobby because of the adverse weather conditions. Although tenants from the upper stories had begun evacuating prior to the fire department's arrival, some tenants were trapped in their apartments on the third floor and had to be rescued by firefighters

Extinguishment of the fire required approximately 25 minutes. Total evacuation of the building took approximately 90 minutes. Four elderly residents died as a result of smoke inhalation; 135 tenants were evacuated and transported to an emergency shelter. In addition to the facilities, 10 residents, 2 firefighters, and 4 police officers were injured. Eighty-eight fire, police, and rescue personnel were required to bring the scene under control.

THE FIRE

On December 14, 1989, at 0214 hours, Roanoke County Fire Dispatch received an automatic fire alarm notice from the Shenandoah Retirement Center at 5300 Hawthorne Road. At 0216 hours, Companies 5 and 10 were alerted by County Dispatch, their volunteer alert system.

A sheriff's deputy, who was in the vicinity of the center, heard the call and responded to the center. Upon her arrival at the scene, she observed flames coming from a third floor window located at the front of the building. She reported the situation to the fire dispatcher who immediately alerted additional fire units. The deputy entered the building through the front entrance and began assisting elderly residents who were already in the evacuation process at the north stairwell.

At approximately 0230 hours, the chief of Company 5 arrived in his private automobile. He entered the building and found the lobby full of residents who had evacuated upper portions of the building. Most were in their nightclothes and were not dressed for the frigid weather conditions outside.

The chief proceeded to the third floor by way of the north stairwell. When he opened the third floor stairwell door, he saw that the floor was completely charged with heavy, black smoke. He began assisting residents to evacuate the building.

At 0234 hours, Wagon 5 arrived with three firefighters. They immediately started a rescue operation, concentrating on the third floor. Their initial entry was by way of the stairwell on the south end of the building. They were successful in removing one or two residents, then back-up arrived a few minutes later.

As more units arrived, rescue began from the stairwells located at both ends of the building. Firefighters experienced great difficulty in gaining entrance to individual apartments, as most apartment doors were locked. Firefighters had obtained and were trying to use a master key, but the dense, black smoke made progress slow. Many third floor residents had already evacuated, but a room-to-room search was necessary.

Meanwhile, firefighting personnel were attempting to locate a standpipe hoseline which was housed in a cabinet at the south end of the third floor hallway. Because of the smoky conditions, they were unable to find the cabinet. At this point, the highrise hose pack carried on the apparatus was connected to the standpipe system located in the stairwell, at the third floor level, and advanced to the room of origin. The firefighters detected some fire in the hallway at the ceiling level in the area of the entrance of Apartment 307. Flashover had already occurred.

Meanwhile, the room-to-room search on the third floor and the evacuation of the entire building were in process. Information from fire dispatch was received by the third floor section commander that a woman was trapped in Apartment 302. Their attempts to locate Apartment 302 were made extremely difficult because of the black soot that covered the apartment numbers. Unfortunately, they were not able to locate the woman in time to save her.

Smoke was beginning to filter to the upper stories of the building by way of stairwells and the doors being opened by people leaving the building. Smoke movement was also aided by a 9-mile-per-hour southeast wind as it came through an open window in Apartment 307.

The incident commander located at the front of the building had already established an emergency medical services (EMS) command post which was in the process of performing triage and transporting some of the residents to hospitals. The emergency service officer had activated the county dispatcher plan and a shelter was established at a nearby roller skating rink. It was staffed by members of the American Red Cross and medical personnel from Roanoke Hospital. Four school buses had been dispatched to the scene to provide transportation of the evacuees to the shelter. Most residents were in nightclothes and needed assistance to walk; many had to be carried a relatively long distance over snow and ice and through an array of fire apparatus in order to be placed aboard buses.

Because of the limited vehicle access afforded at the site, most of the driveway was blocked by fire and medic units. The fire required approximately 25 minutes to extinguish, and total building evacuation and transportation to the shelter took about 90 minutes.

During the overhaul operations, it was determined that the fire originated in Apartment 307 and progressed to the hallway through an open door which did not close behind the occupant as he escaped from the room. His body was found in the hallway a few feet from his apartment door. The body of a female resident was found in Apartment 309. A third body was found in Apartment 304 and a fourth in Apartment 302. The majority of the fire was in the room of origin and in the hallway near this room. Smoke filtered into the other rooms where fatalities were found through spaces under and around the doors.

This fire required the services of approximately 88 fire and rescue personnel, and 35 pieces of equipment.

BUILDING STRUCTURE

This building is located in a mixed-use section of Roanoke County, predominantly residential. It is a 6-1/2-story fire resistant apartment building, built for and occupied by elderly residents. The second floor is licensed by the State as an adult care center for 48 patients who require residential custodial care. The building also includes a ground floor under the south half of the building, which is used for kitchen and related services.

The building is constructed of steel framing protected by reinforced concrete. Floors are 8-inch precast concrete with a concrete topping (very similar to Flexcore precast floor panels). Walls are 8-inch concrete block with a brick facing on the outside.

Floors three through six each have 24 single-occupant apartment units for a total of 96 apartments. Most of the units are one-room apartments, though some units have separate bedrooms (Appendix A). The second floor adult care center is set up with 24 units also, but each houses two occupants who receive special care. The building was designed with a 6-foot-wide main corridor running the length of the building, north to south. Enclosed staircases that discharge to the outside are located at each end. Two elevators are located in the center portion of the building and are used as the lobby area on each floor. The length of the hallways is 270 feet. There are no smoke barrier doors between stairwells.

Each apartment is arranged with one or two sleeping areas, kitchen, bathroom, and a small clothes closet. The living/bedroom areas of each unit have 1/4-inch prefinished wood paneling on wood studs -- not UL rated.

CODES

The building was constructed in 1970 under a local building code which did not require sprinkler systems or smoke detectors. The building was modified in 1976 following a fatal fire after which solid core doors and a hallway smoke detection system were installed. The building is considered to be in compliance with existing building codes and is not subject to fire protection upgrading.

The last inspection by the Roanoke County fire officials was approximately 18 months prior to the fire. This inspection involved only the common areas. The second floor is inspected annually by State authorities because it is a licensed adult care center. This inspection does not involve any other portions of the building.

Examination of a few of the units revealed that some occupants are literally packing their residences with furniture and belongings, and are utilizing multigang electrical adapters for numerous electrical appliances. While authority is not granted to the county to enter and inspect living units, this does not prevent building management from inspecting these areas.

FIRE PROTECTION

The building is equipped with an 8-inch standpipe system connected to the city water main, with hose outlets located in cabinets at each end of the main corridor and the elevator core area. Two-and-one-half inch outlets for fire department use are provided in each stairwell. The trash chute and the trash collection room on the first floor are sprinklered.

The building is equipped with an automatic fire alarm system with pull stations located near stairwell doors on each floor. Smoke detection for hallways is also connected to this system. Battery-operated emergency lights are located in each floor hallway.

Emergency vehicle access is considered very poor, as space for vehicles near the building is extremely limited.

Two fire hydrants are installed on the property. Only one, however, is accessible by roadway.

Each room is provided with a pull switch connected to a light located in the first floor office, in the event assistance is needed. The building does not contain a sprinkler system. Individual apartments are equipped with smoke detectors that are battery operated.

Water main pressure in this area is between 50 to 60 lbs.

ORIGIN AND SPREAD OF FIRE AND SMOKE

The fire originated in Apartment 307 and is believed to the result of an electrical overload which caused ignition of wood paneling. This particular one-room unit had 17 appliances of various types connected to the electrical receptacles provided for the unit (see Appendix B).

The wood paneling began to burn rapidly, generating tremendous heat and smoke. Because the room was not equipped with an automatic sprinkler system, the fire burned unchecked.

The occupant of this unit was confined to a wheelchair. Upon his exit from the room to the corridor, his apartment door remained open, allowing smoke and heat to escape to the hall, setting off the smoke detector located near his unit in the hallway.

Smoke very quickly filled the entire third floor. As residents on this floor began evacuating, smoke entered the stairwells and spread to the upper floors of the building. Flames eventually broke through the window in the room of origin, which is located on the front of the building. There was a southeast wind of approximately 9 miles per hour which helped spread the smoke.

Fire department response required 18 minutes, a factor which undoubtedly played a role in the extent of smoke spread.

The fire department estimates that flashover occurred between 3 to 5 minutes from ignition of the paneling.

FIRE DEPARTMENT

Roanoke County Fire and Rescue Services operates ten fire stations and three medic units with a paid staff of 45 persons. Working hours are from 7 a.m. to 5 p.m., Monday through Friday. All other times these fire stations are operated by a volunteer force. The volunteers respond from home or work. The stations are not staffed after 5 p.m. Normal response time is from 4 to 7 minutes. At present, there does not exist an automatic response agreement with its neighbor, Roanoke City.

This particular incident occurred at 0214 hours, which required that the volunteers be notified at their homes. The temperature was 14 degrees Fahrenheit, with snow and icy road conditions. Response times from the nearest fire station to the fire scene is approximately 5 minutes. The reason it took 18 minutes to respond to this fire (other than the fact that adverse weather conditions contributed somewhat to delay) is not known at this time; however, the county administrator has organized a task force composed of fire service representatives and other county officials to determine the cause of the delay and to recommend corrective action. It is believed that the early morning hour of the fire (extra time may have been required to rouse the volunteers from sleep), and a recent spate of false alarms that repeatedly had called the volunteers to this facility may both have contributed.

BUILDING FIRE HISTORY

This building, since becoming occupied in 1972, has experienced three fatal fire incidents. The first occurred in March 1976 when an elderly female's nightgown caught on fire as a result of careless smoking. She died as a result. The second fatal fire occurred in June 1979 and again was the result of careless smoking which ignited bed clothing. That fire, fueled by the prefinished wood paneling, very quickly engulfed a second floor room located in the adult care section. Four elderly people died and 12 others were injured. A few weeks before the most recent fire incident of December 14, 1989, there had been numerous calls to the fire department due to tenants cooking in their rooms and setting off hallway smoke detectors. Repeated call-outs to the building for these false alarms may have created a sense of lower urgency among the volunteers called to respond in the middle of a freezing night.

BUILDING EVACUATION PLAN

There were three staff members working in the building the night of the fire -- two nursing assistants located on the second floor adult care section and a building night manager. Immediately upon notification of the fire by way of the alarm system, the nursing assistants began evacuating the second floor, moving the people to the first floor lobby area via the stairwells. At the same time, the night manager began evacuating the rest of the building. The first floor lobby area was selected because of the adverse weather conditions and the frailty of the residents.

By the time the first firefighting unit arrived on the scene, the entire second floor had been evacuated and the remaining tenants were proceeding down the two stairwells. Building staff reacted to and followed established procedures in a very proficient manner, which undoubtedly saved lives and prevented further injuries.

IMPLEMENTING THE EVACUATION AND SHELTER MANAGEMENT PLAN

The first-arriving assistant chief established a command post in front of the building and implemented the department's fire incident command procedures. As additional fire service personnel arrived, interior section commanders were assigned to firefighting, rescue, and evacuation procedures. An EMS command post was established, which instituted preliminary triage at the scene. The fire department emergency coordinator activated the emergency evacuation and shelter management plan. School buses were dispatched to the scene and residents were transported to the emergency shelter, which was a skating rink located close by. The American Red Cross disaster teams and medical personnel from the Roanoke Hospital reported to the shelter and provided blankets, bedding, food, and medical services. One hundred thirty-five residents were evacuated and transported to the shelter; ten were transported to the hospital for minor treatment. News media were requested to broadcast information regarding the shelter so that family members would know where to come to help care for the residents. By noon of the same day, all but eight of the evacuees had been called for by family members.

Roanoke County fire and rescue services frequently practice and update their emergency disaster plans. The smoothness and effectiveness in implementing this plan in the Shenandoah fire, under adverse conditions, is truly commendable. Their success can in large part be attributed to their foresight in developing a plan and practicing it.

FATALITIES

Four elderly residents from the third floor died as a result of smoke inhalation -- three women and one man, ranging in age from 79 to 87 years. The body of the resident in Apartment 307, (area of origin), who was confined to a wheelchair, was found in the hall a few feet south of his apartment door. One female victim of Apartment 309 was found in her unit on the floor next to her bed. The third victim (female) was found in Apartment 302. The fourth victim (female) was found in Apartment 304, between the rear of the television and the window. This person was actually having a telephone conversation with the fire dispatcher while fire operations were taking place in the building. She was provided with instructions by the dispatcher regarding covering her nose and putting towels under the apartment door. Meanwhile, the dispatcher was advising firefighters on the scene of the woman's situation and her location. Because of heavy smoke conditions and locked doors, rescuers were unable to locate her in time to save her (see Appendix A).

INJURIES

Building Occupants -- Ten building residents sustained minor injuries due to smoke inhalation or stress related to the fire incident. These people were transported to the hospital and released within 24 hours.

Firefighters -- Two firefighters were injured, one as a result of smoke inhalation and the second as a result of a twisted knee. Both were treated and released from the hospital.

Law Enforcement Personnel -- Three police officers and one sheriff's deputy, who were assisting in the rescue and evacuation effort, were treated for smoke inhalation and related causes. All were treated and released shortly thereafter from the hospital.

DAMAGE ASSESSMENT

Fire damage was mostly confined to the room of origin and hallway at the south end of the third floor. Fire completely gutted the room of origin. There was heavy smoke and water damage to the second floor; moderate smoke damage to the remaining upper portions of the building.

There was no significant structural damage. This was due to the sound construction of the building. Apartment separation walls were constructed of masonry materials.

Examination of the apartment door of the room of origin revealed that the door was open, which allowed the fire and smoke to escape to the hallways. There is some evidence that the self-closing device installed on the door had been adjusted to afford the least resistance possible as the occupant passed through in his wheelchair. This, perhaps, accounts for the door not closing. Also, since the apartment occupant was confined to a wheelchair, he probably found it more suitable for his coming and going through the door.

Dollar loss has not yet been determined, but is expected to be several hundred thousand dollars.

LESSONS LEARNED

1. **Regardless of when they were constructed, multiple occupancy institutional buildings and retirement homes should be subjected to current fire codes regarding installation of fire protection equipment.**

    This fire experience is further testimony to the urgent need for such action. The installation of a sprinkler system, coupled with a well-designed smoke detection system, would probably have reduced, and possibly eliminated, this tragic loss of life.

    When dealing with large numbers of frail and bedridden people, evacuation may not be a viable alternative. This further illustrates the essential need for superior automatic fire suppression and detection capabilities.

2. **Regular fire prevention inspection by both the fire department and building management is absolutely essential in retirement homes.**

    Building management is in an excellent position to augment the local fire departments in conducting these inspections, as they are permitted to enter each apartment unit where perhaps the fire department cannot.

3. **Well-designed emergency procedures and regular employee training contribute to occupant safety and loss control.**

    The Shenandoah Retirement Home fire once again demonstrates the value of developing and implementing a well-designed emergency procedures program. The programs at these facilities are excellent. They are well-designed, clearly documented, and practiced on a monthly basis. This training was evident the night of the fire when staff immediately began closing doors to impede the spread of fire and smoke and began evacuating building residents.

4. **Attention must be given to wall treatments and fuel loading in individual apartments in retirement homes.**

    The prefinished wood paneling, which was used extensively at the Shenandoah Homes facility, and heavy fuel-loading from the furniture and possessions of elderly residents were the main culprits in the rapid flame and smoke spread.

5. **The fire service must be an integral part of the building plans review team for all municipalities.**

    The emergency vehicle access designed for the Shenandoah Homes building is inaccessible at best. It only affords vehicle access to a very small portion of the structure and to one fire hydrant. This problem was most apparent when fire apparatus had to be placed on a snow-covered lawn in order to reach the building and to use the fire hydrant. A complete review of other existing structures that present similar problems should be taken without delay.

6. **Effective incident command plans contribute to the control of life and property loss.**

This fire again shows the absolute need to design and implement effective fire scene management plans. The management of this fire and the subsequent evacuation procedures that were utilized undoubtedly prevented further loss of life and injury.

7. **Mutual aid response should be as automatic as possible.**

The delayed response that was experienced at the Shenandoah Homes fire, regardless of the reasons for it, brings to the forefront the essential need to have immediate, available back-up. A staffed company from the City of Roanoke was only 5 minutes away from Shenandoah Homes. The dispatching of the nearest fire equipment to the scene of an emergency not only affords the citizens an added degree of protection but has, in many jurisdictions, become a viable alternative to budgetary constraints.

As this report was going to publication, the City and County of Roanoke have implemented a joint response policy for nursing and retirement homes located near the city/county border. This policy calls for a full response from both city and county units.

Supplemental Report:
Three-Fatality Fire in Sixteen-Story Apartment Building for Senior Citizens
Watertown, New York

A December 15, 1989, fire at the Midtown Towers, a 16-story seniors apartment building killed 3 residents and injured approximately 70 others. The arson fire was started about 0030 in a community room on the ground floor.

The building was constructed under a Federal housing program for the elderly and is similar to many throughout the United States. The building was equipped with standpipes, and a corridor smoke detection system was added in 1977. The corridor smoke detectors were installed on all floors except the ground floor and mechanical penthouse. The alarm system was connected directly to fire dispatch. Single station detectors were located in each apartment and the ground floor area.

The Watertown Fire Department received the alarm at 0036 from activated smoke detectors on the sixth and seventh floors. Shortly thereafter, they received a phone call from a building occupant confirming the fire. On arrival, firefighters found a freely burning fire involving the first floor community room and kitchen area. Windows on the first floor had failed.

Smoke conditions were worse from the seventh floor up. The principal means of smoke travel was thought to be vertical shafts (bathroom and kitchen exhausts, corridor ventilation, and elevators). Approximately 40 people were removed by aerials. Fifty-seven people were transported to hospitals by ambulance. About another 13 were taken by private automobiles. About 100 firefighters from the city and surrounding communities responded to the fire. The fatalities were located on the eleventh and twelfth floors.

APPENDICES

A. Shenandoah Retirement Home Third-Floor Plan and Location of Fatalities

B. Room of Fire Origin

C. Units Used at the Fire

D. Fireground Diagram Showing Positions of Fire Units at Scene

E. Photographs

APPENDIX A

Shenandoah Retirement Home
Third-Floor Plan and Location of Fatalities

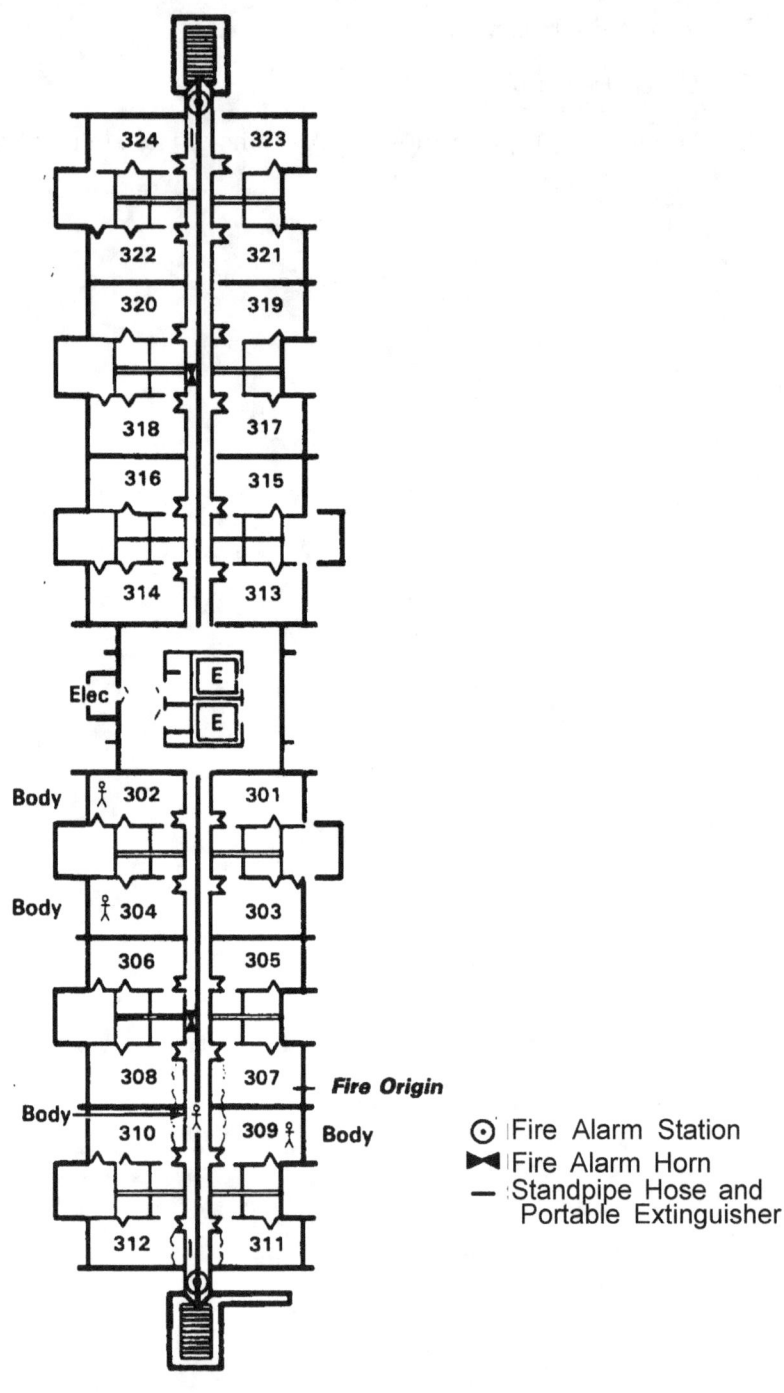

Fire Origin

⊙ Fire Alarm Station
⋈ Fire Alarm Horn
— Standpipe Hose and
    Portable Extinguisher

772-4-9-90-4
R8-8-90

APPENDIX B

Room of Fire Origin

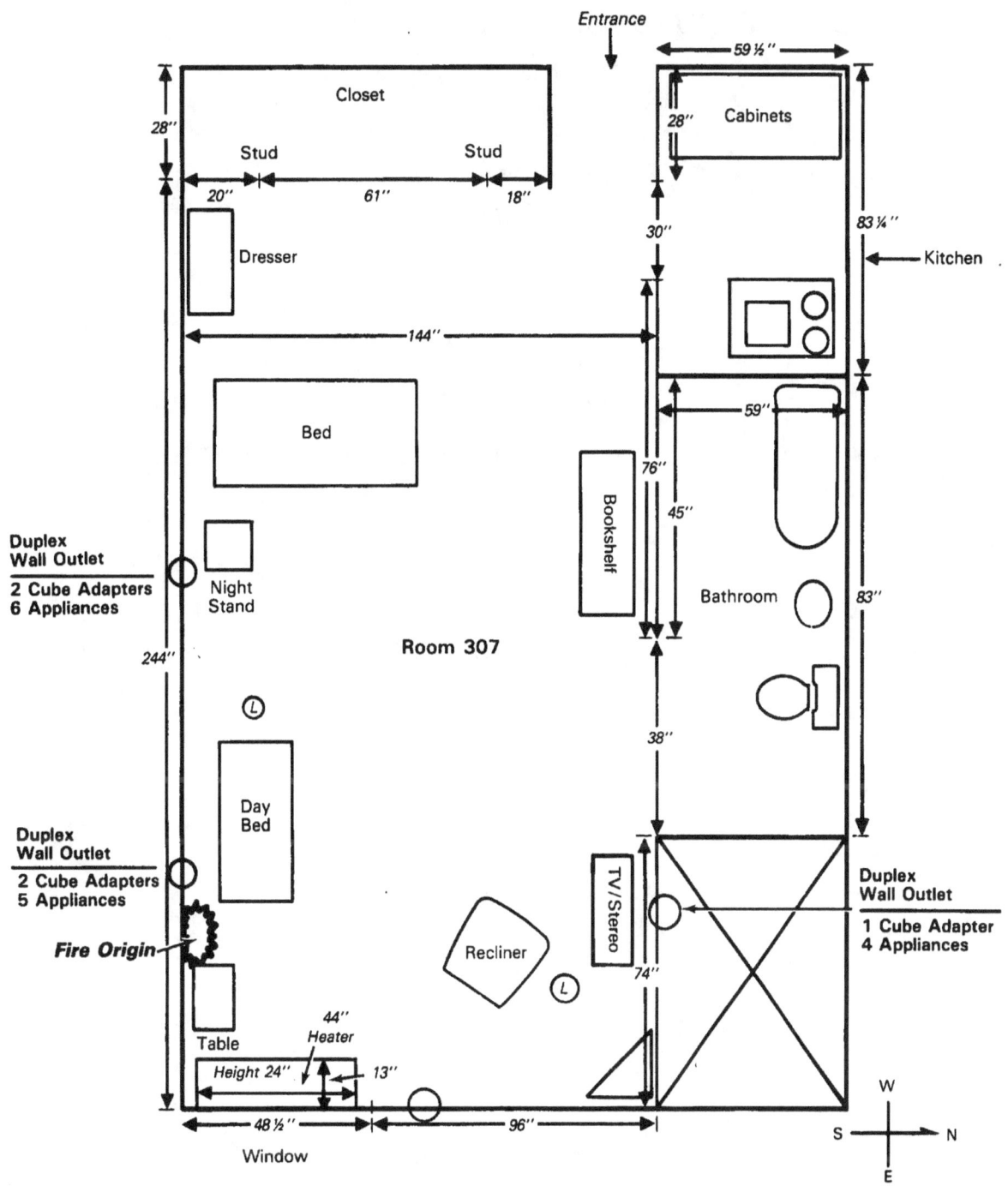

772-4-9-90-3

APPENDIX C

Units Used at the Fire

Equipment:

7 Engines

2 Trucks

3 Heavy Rescue

16 Rescue Units

1 Air Wagon

1 Utility Unit

1 Brush Unit

4 Cars

88 Fire and Rescue Personnel

NOTE: Above includes 3 medic units, 1 crash truck, and 10 staff from Roanoke City Fire Department.

APPENDIX D

Fireground Diagram Showing Position of Fire Units at Scene

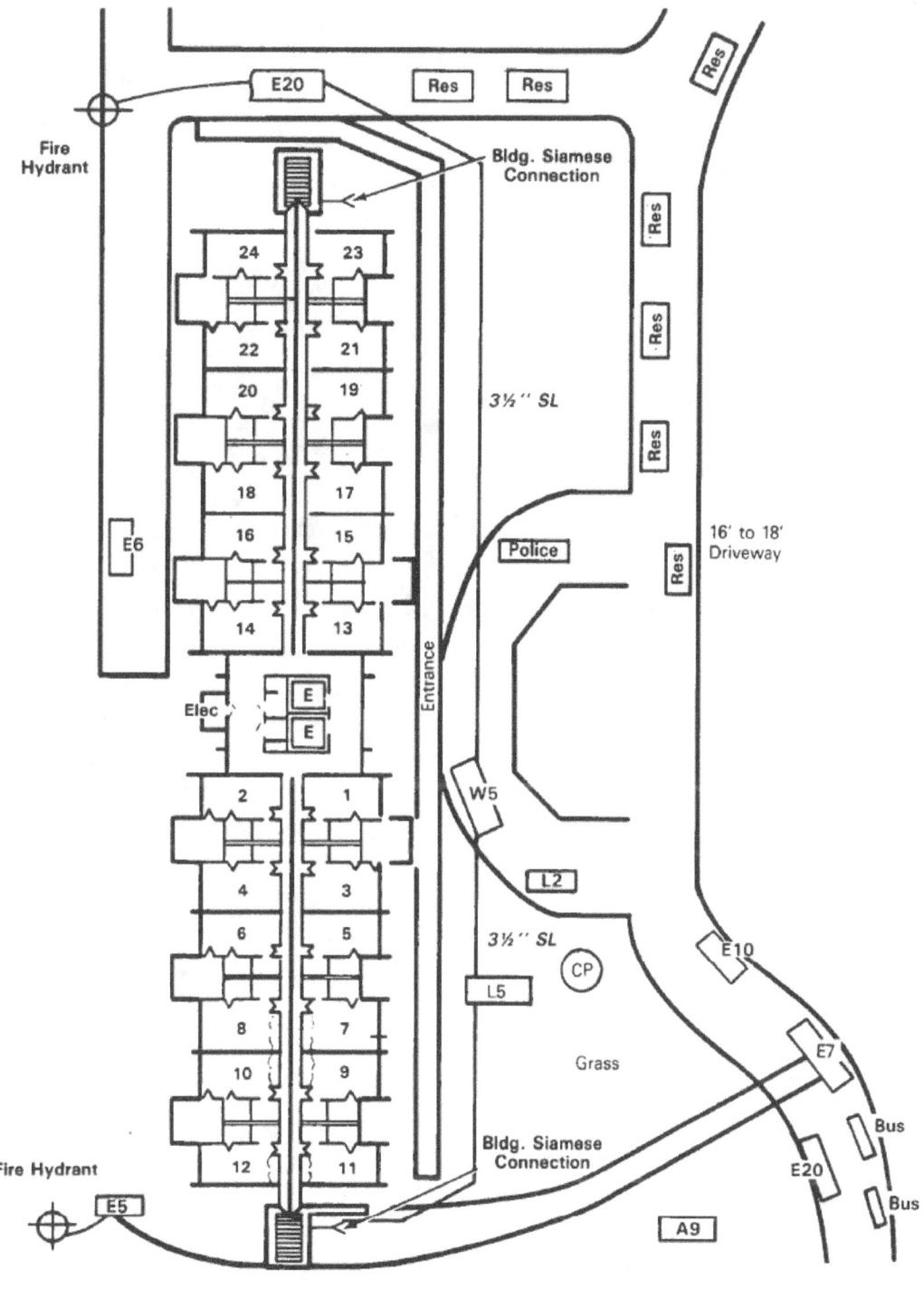

772-4-9-90-5
R8-8-90

APPENDIX E

Photographs

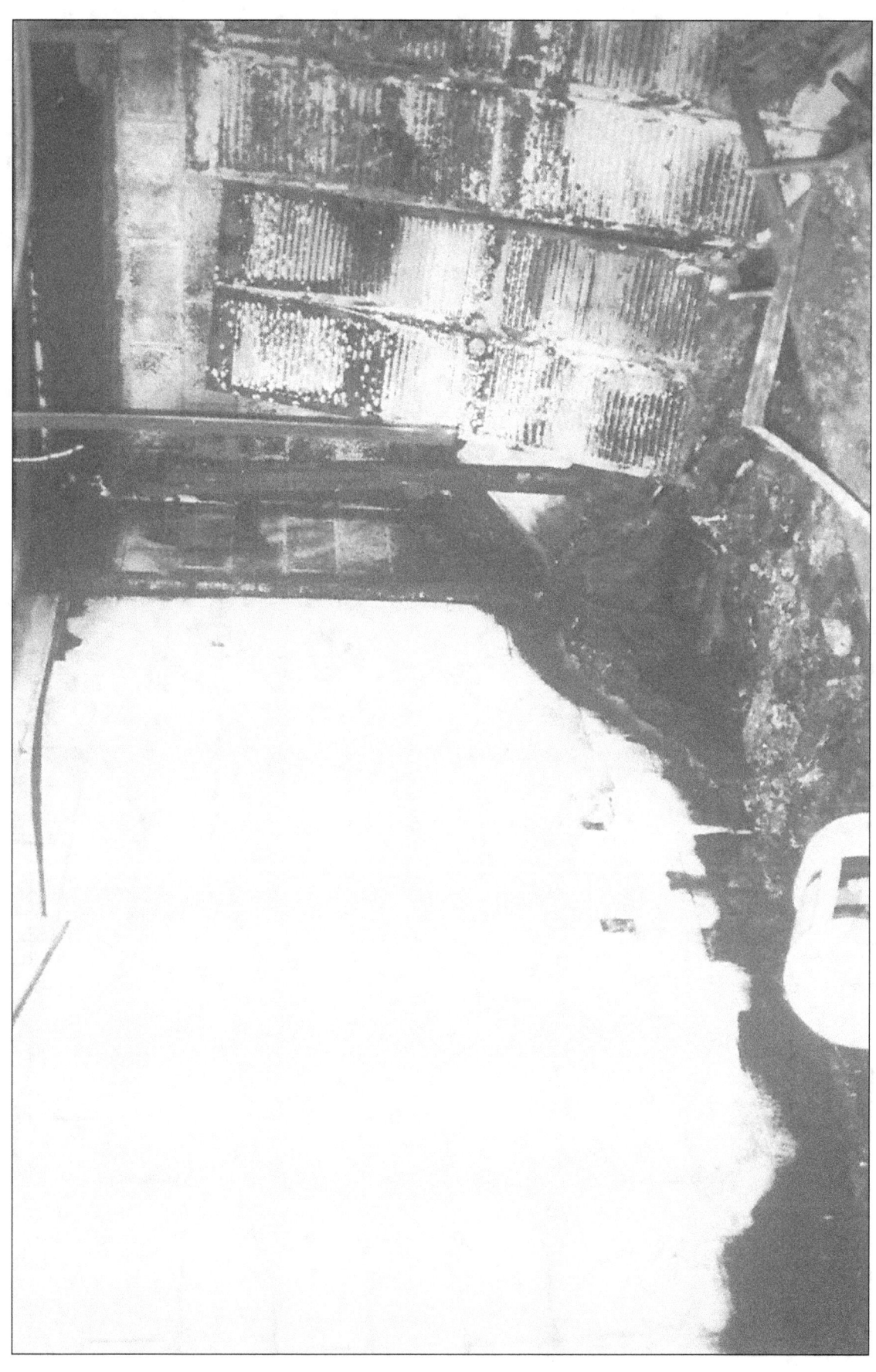

www.ingramcontent.com/pod-product-compliance
Lightning Source LLC
Chambersburg PA
CDIIW081420170526
45166CB00010B/3414